LOOKING AT SCIENCE 3

Changes

David Fielding

Basil Blackwell

© 1984 David Fielding

All rights reserved. No part of this publication may be reproduced, stored in a retrieval system, or transmitted in any form or by any means, electronic, mechanical, photocopying, recording or otherwise, without prior permission of Basil Blackwell Publisher Limited.

First published 1984

Published by Basil Blackwell Limited
108 Cowley Road
Oxford OX4 1JF

ISBN 0 631 91370 X

Printed in Hong Kong

Topic symbols

 This work is about air and water.

 This work is about animal life.

 This work is about electricity and magnetism.

 This work is about light and dark.

 This work is about mechanics.

 This work is about plant life.

 This work is about weather and climate.

Look for the symbols in the other books in the series. There is more work about these things in the other books.

Contents

A word to teachers and parents 5

Part 1 Change in nature

Great oaks from little acorns grow 6
How plants make seeds which grow into new plants. How the seeds travel to the soil.

Sowing the seed 8
How seeds start to change into plants.

Rotting 10
How dead plants rot and are changed into food for new plants.

Hot and cold 12
What climates are. Why different parts of the world have different climates.

Summer and winter 14
Why we have the change from summer to winter and back each year.

The changing seasons 16
How plants and animals alter to cope with the change to winter.

Growing things and the seasons 18
A look at what happens in winter, spring, summer and autumn.

Part 2 Changes in air and water

Fire 20
How a fire can be put out by changing the air round it. How a fire alters the air around it.

Light as air 22
How air changes when it is heated. How this change can be useful.

Now you see it, now you don't 24
How water can change to vapour on a warm day, and change back when it gets cold.

Held on 26
What air pressure is, how it changes, and why.

Deep down 28
What water pressure is, how it changes, and why.

Unsafe to drink 30
Some things dissolve when they enter water. Dissolved things can make water unsafe to drink.

Changes in air and water 32
The atmosphere and the sea contain many changes in pressure, temperature and composition.

Part 3 Changes seen and unseen

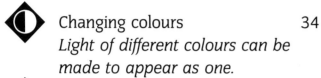
Changing colours   34
Light of different colours can be made to appear as one.

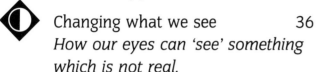
Changing what we see   36
How our eyes can 'see' something which is not real.

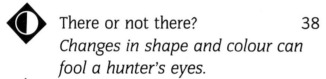
There or not there?   38
Changes in shape and colour can fool a hunter's eyes.

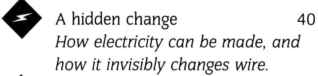
A hidden change   40
How electricity can be made, and how it invisibly changes wire.

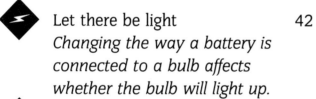
Let there be light   42
Changing the way a battery is connected to a bulb affects whether the bulb will light up.

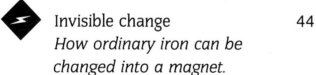
Invisible change   44
How ordinary iron can be changed into a magnet.

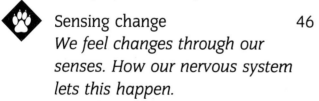
Sensing change   46
We feel changes through our senses. How our nervous system lets this happen.

New words   48

Acknowledgements

The Australian Information Service, London   15 (4)
Barry Angell   17 (3), (4)
B & B Photographs   7 (2), (3), (4)
Biofotos   39 (2), cover
Janet and Colin Bord   15 (3), 37 (3)
Cameron Balloons Ltd   22 (1), (4)
The Electricity Council   40 (1)
Ever Ready Ltd   41 (3), 43 (3)
ICI Plant Protection Division   31 (3)
Keystone Press Agency Ltd   24 (2)
London Fire Brigade   21 (4)
Milk Marketing Board   31 (4)
Ministry of Defence   39 (4)
Paul Newton   39 (3)
Oxford Mail and Times   20 (1)
Ken Pilsbury   13 (3)
Popperfoto   13 (5)
Rentokil Ltd   11 (2), (3)
Elizabeth Skillman   35 (4)
Rebecca Skillman   27 (4), (5)
Flip Schulke/Seaphot Limited: Planet Earth Pictures   29 (4)
Harry Smith Horticultural Photographic Collection   11 (2)
Thomson Holidays   32 (2)
Topham Picture Library   31 (2), 25 (4)
ZEFA   33 (3), 47 (4)

Illustrations by Michael Stringer (colour)
and David Fielding (black and white)
Design by Indent, Reading

A word to teachers and parents

Changes is one of five books in the *Looking at Science* series. The series has been designed to do two things:

- It gives children a solid body of knowledge in natural and physical science.
- It begins to introduce the nature of scientific enquiry.

These two elements are developed side by side through the books.

Each double page covers a particular area for study. The left-hand page outlines an activity to perform, while the right hand page gives information connected with it.

The activities are introduced with the symbol ❤, and cover experimentation, observation and recording. A list of all the equipment needed for the experiments is given near the beginning of each spread.

Each book also contains another kind of double page, which is purely factual, spaced at regular intervals. These pages draw together the themes of the preceding pages.

These pages can be worked through in order. Alternatively, they can be used as source material for topic work. Suggested topic areas are identified, with symbols, in the contents list.

Notes for Book 3

The work in this book explores the theme of 'change' in three areas.

Part 1 Change in nature

This section looks at changes in the cycles of nature. The opening units deal with the beginning of a plant's life, as seed and then seedling. The next unit looks at the other end of its life, as it rots into humus. Two units explain variations in climate in the world, and why the seasons change during the year. The next unit explores how plants and animals respond to these changes. The final unit links all this work together by giving a comprehensive picture of the changes that happen in nature in the course of a year.

Part 2 Changes in air and water

This section turns from changes on land to changes that occur in air and water. The first unit shows how fires need air, yet change air by using up its oxygen. The next unit shows how air changes when it is heated. The third unit explains the changes of evaporation and condensation in the air. Two units explore the idea of pressure and how air and water behave when their pressures change. A unit about pollution shows how water can be changed from something safe to something unsafe when things dissolve in it. The final unit puts all these ideas together in a comprehensive view of the world's atmosphere and seas.

Part 3 Changes seen and unseen

This section shows how our senses can be misled by some changes. The first three units explore visual changes: changes in colour; changes in position; changes in patterns. They include work on illusions and camouflage. The next three units explore the idea of changes which our senses cannot pick up directly. They look at simple electricity and magnetism. They explain how electricity can be made and used, and how some things can be magnetised. The final unit looks at the body's mechanism for sensing change, the nervous system, and how it works.

Part 1 Change in nature

Great oaks from little acorns grow

In autumn, most trees lose their leaves and spread their seeds. They do this in different ways.

You will need cardboard, Plasticine, scissors, a knife, a tomato, a margarine tub and compost

♥ Experiment: Making a flying seed

Use cardboard and Plasticine to make a copy of a sycamore seed. Drop your model from as high up as you can safely manage. Watch how its 'wings' make it spiral down. Drop it again, but this time ask a friend to make a wind by flapping some cardboard. See how far your sycamore seed is blown.

Picture 1 How to make a model flying seed

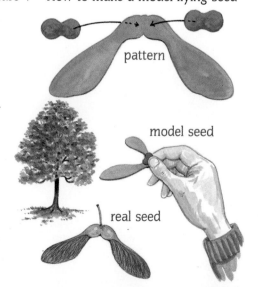

Picture 2 Starting the experiment

♥ Experiment: Growing seed

Cut a tomato in half. Leave one half aside to draw. Look closely at the other half. Find the seeds in it, and take them out. Put some moist compost in a margarine tub. Press the seeds just beneath the surface, and seal the tub. Put it somewhere dark and warm. Open it every few days for two weeks. See if the seeds grow.

Picture 3　A dandelion 'clock'

Picture 4　The seeds of this berry are carried by birds

❤ *Record*

1 Describe your model sycamore seed and how it flew. Do you think that real sycamore seeds can travel far? Why do you think that seeds need to travel?

2 Explain what your fruit was like inside. Draw its inside. Say how you planted the seeds. How might these seeds normally reach the soil?

Seeds that fly

Seeds need to travel away from their mother plant, to find fresh soil to grow in. Some seeds are blown by the wind. Most flowers have this kind of seed. So does the sycamore tree. Seeds that travel on the wind need to be very light, or have 'wings'.

Seeds in fruit

Fruits with seeds inside them fall to the ground. Animals eat them. The soft fruit feeds the animal. The hard seed is no good to it. It goes right through the body. It comes out with the dung and falls to the ground. The animal may have taken it a long way from the mother plant.

Seeds that are carried

Some seeds have a spiky covering. They catch on the coats of animals and are carried to fresh ground. After a time, they fall off. The outside covering dries in the sun and splits open. Then the seed is free to grow.

❤ *To write*

1 Why do plants make so many seeds?
2 Why are some seeds hidden inside tasty fruit?
3 What would happen if seeds did not travel?

❤ *Something to do*

Collect different seeds. Label them, and find pictures of the plants they come from. Find out how each seed travels to reach the soil.

Picture 5　How do you think this seed travelled?

Sowing the seed

Picture 1   Prepare the experiment like this

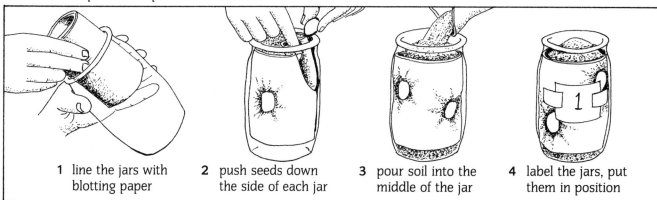

1 line the jars with blotting paper
2 push seeds down the side of each jar
3 pour soil into the middle of the jar
4 label the jars, put them in position

Seeds grow better in some places than others.

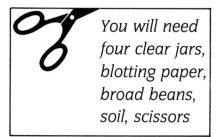

You will need four clear jars, blotting paper, broad beans, soil, scissors

♥ *Experiment: To see what makes seeds germinate* (Germinate means 'begin to grow'.)

Line four clear jars with blotting paper. Place one or two broad beans in each jar, between the blotting paper and the side of the jar. Fill each jar with soil. Number each jar. You have begun four experiments. So far, they are the same.

1 Water jar number one and put it in a warm, dark place. Keep the soil moist.

2 Water jar two, and put it in a cold, dark place. A refrigerator would be ideal. Keep the soil moist.

3 Put jar number three in daylight. Keep the soil dry.

4 Put jar number four in a dark place. Keep the soil dry.

Check the jars every day, to see if the beans germinate.

♥ *Record*

Describe what you have done. Draw how you arranged the beans in the jars. Make this chart to record the progress of each bean:

	jar one	jar two	jar three	jar four
day one				
day two				
day three				
day four				
day five				

Do you think that all of the beans will germinate? Which beans will germinate first? Why? You put each jar into different conditions. Do the conditions matter? What other conditions could you try?

Germination

Soil
Seeds need soil in order to grow into plants. Good soil is called *loam*. It has plenty of goodness (called *humus*) in it. Humus is made from rotted plants and dung.

Moisture
Seeds need moisture in order to germinate. Water is able to soak into most seeds. This makes the seed swell, split its coat and begin to grow. Loam keeps a good amount of moisture in it.

Air
Seeds need air, even when they are underground. Loam has plenty of air in it

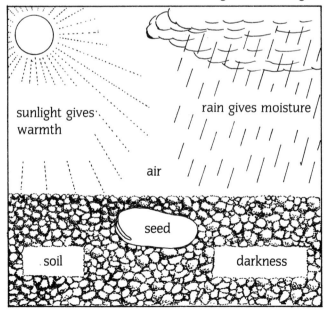
Picture 2 A seed needs different things to make it grow

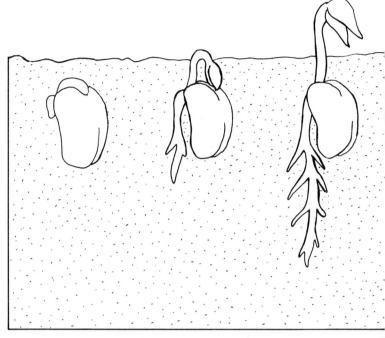
Picture 3 This shows how a seed grows, and makes leaves and roots

because soil creatures keep the soil loose. Air gets between the loose lumps of soil.

Darkness
Seeds are best buried in the soil. Then birds cannot eat them and the sun cannot scorch them. They grow best in the darkness below the surface. They grow up strongly to find the light. At the same time, roots grow downwards to find water.

Warmth
Seeds need some warmth in order to grow. If they are planted in cool shade, they do not grow well.

Picture 3 shows how a seed germinates and starts to grow.

❤ To write
1 What kind of soil is loam?
2 What is humus?
3 Why is it better to bury seeds than leave them on the soil?

Rotting

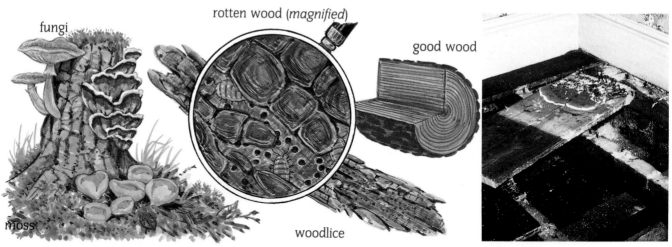

Picture 1   Good wood (*right*) and rotten wood (*left*). Inset – rotten floorboards crumble away

Natural things like wood and plants go rotten when they die.

> *You will need a piece of rotten wood and a piece of good wood, a magnifying glass, a pair of scissors, paper, two margarine tubs, soil, leaves*

Compare a piece of rotten wood with a piece of good wood. See how hard or soft the pieces are. See how the rotten wood will crumble. Spread out the crumbs on paper. Scrape off anything that is growing on the wood. Spread it out on the paper. Look carefully through a magnifying glass. See if you can spot any living creatures.

 Experiment: To see how things rot

Take two containers. Put some fresh leaves in each one. Fill one container with wet soil. Pack it round the leaves, and put on the lid. Look at the leaves every few weeks. You will find a difference between the wet and dry leaves. The ones in wet soil will slowly begin to rot.

♥ Record

Describe what good wood is like, and what rotten wood is like. Say if any tiny plants or animals lived on the rotten wood. Say where the rotten wood came from.

Describe how you set out to make leaves rot. Draw what you did. What do you think makes things go rotten?

Decomposers

When plants and animals die, they *decompose* or rot. They go soft and crumbly. Later, they seem to disappear. They change into other things.

Decomposers change dead plants and animals by eating them. Some decomposers can be seen.

Picture 2 A woodworm grub

You may have seen mites on your rotten wood. They have been eating it. Most decomposers are too small to see. *Bacteria* are the most important decomposers of all. Bacteria are tiny. They can be dangerous to us, because some cause diseases. Some decomposers are plants. The one in Picture 4 is a *fungus*. It grows on dead things and gets its food from them.

Picture 3 The leaves and grass in this compost heap are rotting into humus

Picture 4 Decomposers live on dead plants and animals

Rotting is important

Decomposers change dead things by eating them. They turn the dead things into stuff called humus. Humus is full of the kinds of things that plants use as food. Soil with plenty of humus in it grows strong plants. When dead plants and animals have rotted, they become food to help new things grow.

❤ To write

1 Humus is the remains of organic things. True or false?
2 What are decomposers?
3 What would happen if dead things did not rot?

Hot and cold

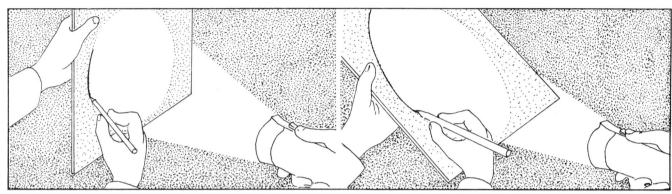

Picture 1 Experimenting with light on different surfaces

Some parts of the world are colder than others.

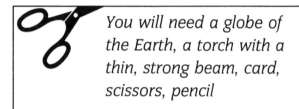

You will need a globe of the Earth, a torch with a thin, strong beam, card, scissors, pencil

❤ Experiment: To show how light spreads

Work in a dark place. Hold a torch level, and shine it at an upright square of card. Draw round the circle of light made by the torch. Then hold another square of card in front of the torch, but slope this card backwards. See how the area of light spreads. Draw round it. Compare the two shapes you have drawn.

❤ Experiment: To show how light spreads on Earth

Put a globe in front of the torch. Find the north and south poles, and the *equator*. Hold the torch level, and shine it at the equator. See how the light makes a circle. Lift the torch and shine it at the north pole. See how the light spreads out, like it did on your sloping card.

❤ Record

Describe the experiments, and draw what you did. Explain how the light made different shapes on the cards. Say where on the globe the light shone most strongly. Where did it shine most weakly? If the torch is the sun, where are the hottest parts of the world? Where are the coldest? Why?

Picture 2 This experiment shows how the sun lights the Earth

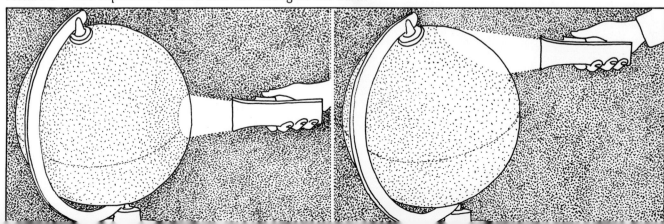

Climates

We call rain, cold, wind and so on by one word. The word is 'weather'. We talk about the weather today, or this week. *Climate* is a word to describe what the weather is like all the year round.

Picture 3 Stormy weather at the seaside

Polar climate

At the poles, the Earth's surface curves away from the sun. It curves away like your card sloped away from the torch. This makes sunlight spread out over a big area. The sunlight is too weak to warm the ground. Ice builds up, and the poles get very cold. They have a polar climate.

Tropical climate

The equator does not curve away from the sun as much as the poles. Like the upright card, it collects strong light which is not spread out. The strong sunlight makes the equator hot. The equator has a tropical climate.

Picture 4 An icy desert in the Arctic circle (*top*). The sandy desert of Death Valley, USA (*bottom*).

Temperate climates

Countries in between have a temperate climate. Their summers are not too hot, and their winters are not too cold.

Each kind of climate shades gradually into another one.

❤ *To write*

1 Why do the poles have a cold climate?
2 Why is it hottest when the sun is high in the sky?
3 What kind of a climate do we have?

13

Summer and winter

What makes the seasons change?

You will need a globe of the Earth, Plasticine, a powerful torch

♥ Experiment: To show why days are longer in summer than winter

Look at a globe of the Earth. Find on it a country quite near to the north pole. Mark the country with Plasticine. Find a dark place. Rest a torch on a box so that it shines at the globe. Hold the globe so that the north and south poles are in an upright line. See how the torch lights half of the globe. Spin the globe slowly, and watch the country you marked. See how its 'days' and 'nights' are equally long. Next, tilt the north pole towards the light. Spin the globe again. Watch your country pass through day and night. See how it spends longer in daylight than in darkness. Tilt the north pole away from the light. Spin the globe and watch your country. See how its day is now much shorter than its night.

Picture 1   Why the length of day changes from winter to summer

♥ Record

Draw how you held the globe in the light, and explain what you did. Say how tilting the globe gave long days and short nights to some places. Might the Earth tilt like your globe? Would this explain why days are longer in summer?

Picture 3 Christmas in Britain
Christmas in Australia

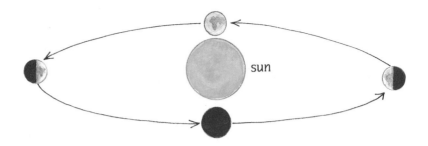

Picture 2

Day and night

The sun and Earth float in space. They are 150 million kilometres apart. The Earth spins around in the sunlight. We go round on it. We spin around into sunlight and darkness in turn, and this is what causes our days and nights. As well as spinning, the Earth moves. It travels in a huge circle around the sun. It goes round the sun once each year.

Summer and winter

As it moves, the Earth is tilted. Sometimes, the north pole is tilted towards the sun. Then, places near the north pole spend more time in sunlight. Days are long and nights are short. They have their summer. Sometimes, the north pole is tilted away from the sun. This brings winter to the north. Northern countries spend less time in sunlight. Days are cold and short, and nights are long.

Summer and winter happen in the south in the same way. The south has winter when the north has summer.

❤ To write

1 It takes the Earth a _____ to travel round the _____
2 Why do summer and winter happen?
3 If the Earth went round the sun twice a year, how many summers and winters would we get each year?

Picture 4 This shows how the Earth turns and why there is winter and summer

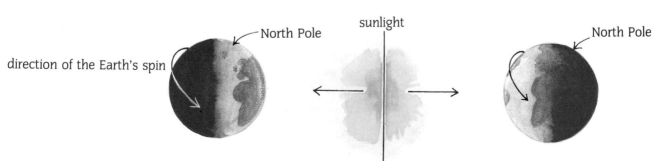

The changing seasons

In autumn, plants and animals get ready for winter. They can tell that winter is coming.

 You will need seedlings, cardboard to make a cover, scissors, sellotape, a plastic tray, a ruler, soil, woodlice

♥ *Experiment: To show that plants change when light changes*

Cover some seedlings so that light reaches them from one direction only. The seedlings will bend towards the light after a day or two. Then move the cover so that light reaches them from another direction. See how they bend in the new direction. (You can grow seedlings from watercress seeds in a couple of days.)

♥ *Experiment: To show how animals notice changes*

Collect some woodlice by turning up old wood or stones. Half-fill a tray with soil. Bury a plastic ruler like a fence across the middle, but leave the soil smooth. Water one side of the tray and cover that side. Put the woodlice on to the dry soil, carefully. After a few minutes, take off the cover. See how the woodlice have all gone to the damp side.

Picture 1   Arrange the experiment like this

Picture 2   Do woodlice like dry or damp soil?

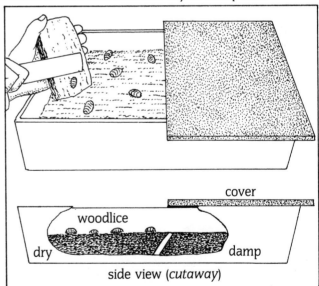

Record

Describe your experiments and draw them. Explain how your plants and animals behaved when things around them changed. What other changes might plants and animals be able to spot? Think up tests of your own to try. How might plants and animals tell when summer changes to autumn?

Sensing changes

We have *senses* that tell us when the weather changes. Our bodies behave differently in hot and cold weather. When we are cold, we shiver. This makes our muscles work, and the movement helps us stay warm. When we are hot, we sweat. This helps to cool us, because as the sweat leaves our bodies it takes heat with it.

Picture 3   A hibernating dormouse

Animals can sense changes

Animals have senses. They can tell when the days are getting shorter and cooler. This makes them change. It makes most birds *migrate*, which means that they fly away to warmer countries. It makes some animals *hibernate* until spring. This means that they go into a deep sleep that is almost like death. It makes some creatures gather extra food. Some animals turn white, to help them hide in snow.

Plants know when change is happening

Plants do not have senses like animals, but scientists think they can tell when days get shorter and cooler. Plants stop growing. Trees lose their leaves. They wait until spring comes round again.

To write

1   How do animals behave when they sense change?
2   What does an animal do when it hibernates?
3   How do plants know when summer is fading?

Picture 4   Swallows gather together before migrating south for the winter

Growing things and the seasons

Spring
Spring comes as the earth tilts into the sunlight. The days grow longer and it gets warmer. Plants and animals sense the improvement. Buds on plants start growing into leaves. Seeds grow warmer in the ground and start to turn into plants. They put out shoots and roots (they germinate). Plants start to make pollen. Birds which have migrated return. Animals waken from hibernation. Many animals have their babies. You can see lambs with the sheep, and foals with horses.

Summer
In summer, the Earth is tilted to face the sun. Days are long and usually warm. Nights are short and mild. Plants grow fruit and flowers. They send pollen to other plants. This fertilises the other plants, and they start to make seeds. Plants grow strongly all through the summer. They use sunlight to make food for themselves. On the farms, the crops grow strongly. At the end of the summer, most crops are harvested.

During the summer, animals bring up their babies. They feed them well to make them strong. They teach them how to look after themselves.

Autumn
In autumn, the Earth tilts so that there is less sunlight. The days get shorter and the nights get longer. It gets cooler. Plants and animals sense the change. Trees lose much of their green stuff or *chlorophyll*. Leaves lose their green colour and fall to the ground. They begin to rot, and make new humus for the soil. This will help the new growth next spring. Fruits and seeds fall to the ground. Animals grow thicker coats and store extra fat in their bodies. Many animals build up a store of food and many birds migrate.

Winter
A country has winter when the Earth tilts it away from the sun. Days are short and dull; nights are long and cold. There are frosts and may be snow. Plants do not grow. Most trees are without leaves. Seeds which fell in autumn lie in the ground but do not germinate. They wait for spring. There is little food for animals, and certainly not enough to feed new animal babies. Most animals stay hidden. They have stored food to help them through the winter. Some of them are hibernating and many birds have migrated.

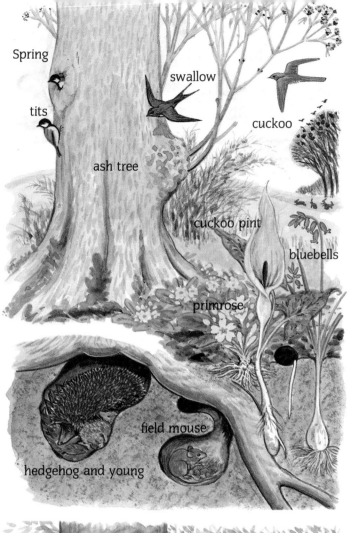

Part 2 Changes in air and water

Fire

Firemen often use water to put out fires, but there are other ways.

Picture 1 Fireman putting out a blaze with water

> You will need a bowl, water, a glass jar, Plasticine, a short candle, matches

♥ Experiment: To stop something burning

Fix four Plasticine 'legs' to the rim of a glass jar. Pour water into a bowl until it is about two centimetres deep. Stand the jar upside down in the water, on its legs. See how far up the water level is inside the jar. Then take the jar out of the water. Stand a short piece of candle in the bowl and light it. Place the jar back in the water over the candle. Watch the candle flame. Watch the level of water inside the jar. Try the experiment a few times. See if the same thing happens each time.

Picture 2 Set the experiment up like this

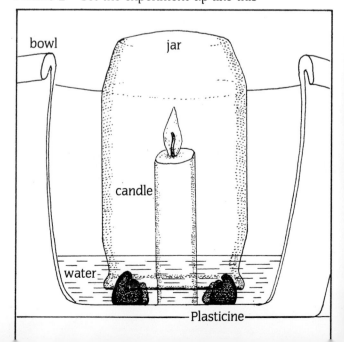

Record

Describe the experiment. Draw pictures to show what happened. Think why the candle went out. Can a flame keep burning in a small place? Can a fire burn without air? Does a flame use up air? Does it use up all the air, or only a part of it? How can you put out a fire without using water?

Fires need air

A flame soon goes out when there is little air. There was only a little air inside your jar, and so the flame soon died. A flame uses up air as it burns. The water showed this when it moved up inside the jar. It moved up to take the place of air that the flame had used.

Oxygen

The water moved only part of the way up inside the jar, because only part of the air had been used. Fire only uses one part of the air. Air contains a gas called *oxygen*. It is oxygen which lets things burn. The candle flame in the jar used up all the oxygen in it, and then went out. The rest of the air was no good to it. Without oxygen, the flame could not burn.

Fighting fires

Fires can be put out by smothering them. This stops air from reaching the fire. Without air, the fire has no oxygen to let it burn, and so it goes out. Firemen use foam and blankets to smother fires.

Picture 3   Why did the candle go out?

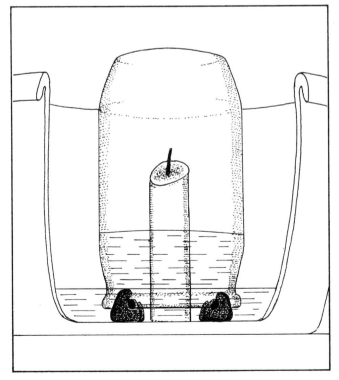

Picture 4   These firemen are using foam

To write

1   Oxygen is a _____. It is present in the _____.
2   What happens to a fire if oxygen is kept away from it?
3   How can foam and blankets put out fires?

Light as air

People fly for miles in huge balloons. Balloons have no wings, propellers or jet engines. There is only air inside them. How do they fly?

> *You will need metal foil, candles, scissors, matches, drawing pin, cotton, a length of wood*

❤ *Experiment: To show how hot air rises*

Cut a disc out of stiff metal foil, ten centimetres across. Cut eight slits in it, spaced evenly around the disc. Each slit must be about three centimetres long. Fold the foil by each slit, as Picture 1 shows. Next, dangle the disc on a thread of cotton, or fix it to some wood with a drawing pin. It must be able to turn easily. Light four candles, and hold your foil in the air above them. See how the foil turns.

❤ *Record*

Describe the experiment and what happened. Use drawings to help. Try to explain why the foil turned. What did the candles do? Would the experiment work without air?

Picture 2   A hot-air balloon

Picture 1   This experiment shows how hot air rises

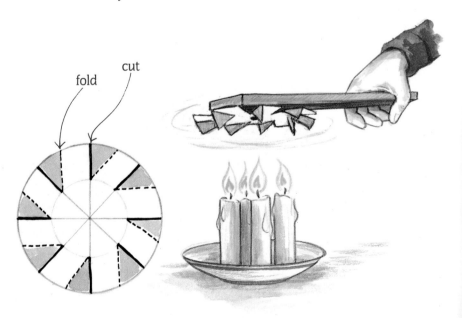

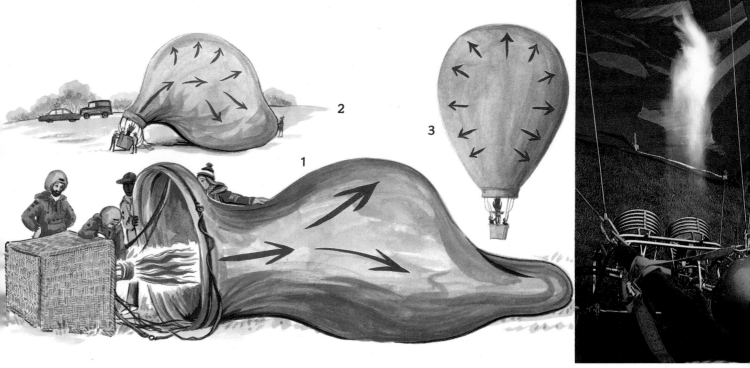

Picture 3   Inflating a hot-air balloon. Inset – the gas-burner for a balloon

Hot air

Hot air made your foil disc spin. The candles heated the air. As this air became hot, it got lighter and rose upwards. Hot air always rises. As it went up, it made a draught past your foil. The draught caught the folded parts of the foil and pushed them round. This made the foil turn.

How a balloon flies

The balloon is laid out flat. A gas burner points into it. The flame makes the air inside hot. The balloon swells, because hot air swells and gets bigger. We say it *expands*. Hot air also gets lighter. So the balloon fills with hot, light air. This light air makes the balloon rise.

Hot air heats rooms

Radiators make the air around them warm. As the air grows warmer, it becomes lighter and rises. It leaves a 'space' where it was. Air which has not been warmed moves over to fill this space. The radiator soon heats this air, too. It also rises. It leaves room which is filled by yet more cool air. As this keeps on happening, the air in the room starts to circle around. This is too gentle for us to notice, but it warms the whole room.

Picture 4   How hot air circulates around a room

❤ To write

1 How does heat change air?
2 How does a hot air balloon fly?
3 What do you think happens to warm air that starts to cool?

Now you see it, now you don't

On a sunny day, rainwater vanishes. What happens to it? Where does it go?

 You will need water, ice, two saucers, clingfilm, a jar

♥ Experiment: To show how water can vanish

Half-fill two saucers with water. Place them on a sunny windowsill. Cover one saucer with clingfilm. Keep a watch on the saucers for a few days. You will find that the water in the open saucer will disappear. It will *evaporate*.

Picture 1

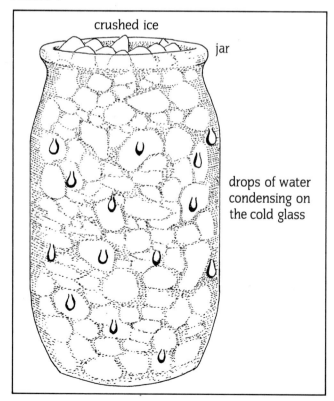

♥ Experiment: To make water appear from thin air

Put some crushed ice into a dry jar. Leave it for a few minutes, and then run your finger around the outside of the jar. See if you find any water there.

♥ Record

Draw pictures of the experiments. Say what you did, and what you found in the second experiment. Where must the water have come from in the second experiment? Does the air have water hidden in it?

Evaporation

When water seems to vanish, it really only changes. It goes into the air as *water*

Picture 2 All the water has evaporated from this reservoir

vapour which cannot be seen. Warmth helps this to happen. When water disappears like this, we say that it *evaporates*. The heat of a sunny day makes water evaporate from wet playgrounds and wet grass.

Condensation

Water vapour which has evaporated can turn back into ordinary water. When air is made colder, the vapour in it turns back into water. We say that the vapour *condenses*. Your jar of ice made the air around it cold. The water vapour in the air condensed on the jar and made it wet.

An imaginary journey

Evaporation and condensation are happening all the time. This is what could happen during one aeroplane journey:

1 The aeroplane is soaked with rain on a cold, wet airport.
2 The wet aeroplane climbs above the clouds. It is in sunlight. It dries. The water evaporates.
3 The aeroplane flies high, into cold air. Its metal gets cold.
4 The aeroplane starts to land. The cold aeroplane sinks back into warm air. It becomes wet as water vapour condenses on its cold metal.
5 The aeroplane lands on a hot airfield. It gets hot. The water evaporates again, and the aeroplane becomes dry.

❤ *To write*

1 What is water vapour?
2 When air is made cold, what happens to any water vapour that is in it?
3 Give examples of evaporation and condensation.

Picture 3   The vapour trails from aeroplanes are streams of condensed water droplets

Picture 4   Find examples of evaporation and condensation happening in this picture

Held on

Picture 1

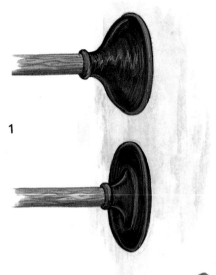

1

2

As a rule, things fall if they are not supported. Does this rule ever change?

 You will need a suction pad, a jar, card, scissors, water

❤ Experiment: Using a suction pad
Hold a suction pad gently against a wall. Let it go and see how it falls. Now press it hard against the wall. It will not fall. It will stick hard.

❤ Experiment: A trick with water
Fill a jar absolutely full with water. Work over a bowl or the sink. Cut a square of card just large enough to cover the top of the jar completely. Lift the jar and hold the card flat over the top of it. Hold it firmly. Slowly turn the jar upside down without letting any water come out at all. When the jar is steady, take your hand away from the card. See if the card will stay in place. Try this a few times.

Record
Describe and draw the experiments. Say what happened each time. Try to think why it happened. What was pushed out when you pressed the suction pad? What was around it as it clung to the wall? Would the experiment work if there were no air?

Air pressure

Your experiments did not change any of nature's rules. The suction pad and water were supported all the time. They were supported by air *pressure*. Above you, there is a huge amount of air, all pressing downwards. We call this pressing the *pressure* of air. We do not notice air pressure because we are used to it. But air pressure is strong.

How a suction pad works

When a suction pad just touches a wall, it has air inside and outside. There is an equal amount of air pressure on both sides. When you push it against the wall, you push out all the air from inside. There is no air pressure left inside to match the pressure outside. The pressure outside keeps the pad pressed flat on the wall.

Picture 2   What makes the suction pad stay on?

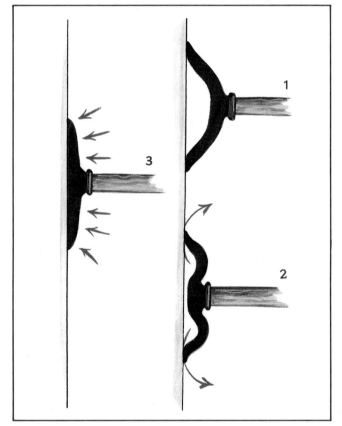

The card and water trick

It was the same with the card. Air kept the card in place. The water pressed downwards on the card. Air pressure pushed upwards on the card. Air pressure won.

Picture 3   How the card and water trick works

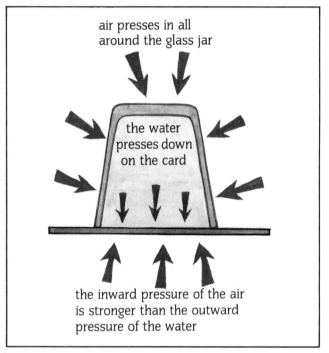

♥ To write

1   Give an example of how we use air pressure.
2   Why will the suction pad fall if the wall is not smooth?
3   Why did the card hold the water in the jar?

Picture 4   When the air pressure in a tyre is low the tyre looks flat (*left*). The pump pushes in more air, the pressure inside gets stronger and fills out the tyre (*right*)

Deep down

A submarine is being attacked. It cannot go up, because the enemy is there. It cannot stay where it is, because depth charges are exploding. Yet the captain is afraid to go deeper because the submarine has sprung a leak. If he goes deeper, will it get worse?

> You will need a squeezy bottle with three holes down the side, Plasticine, water, something for pouring water, a bowl

Experiment: To show how depth affects water

Take a squeezy bottle with three holes down its side. Use Plasticine to plug the holes closed. Leave the top off the bottle.

Picture 1 Water pressure can be stronger than metal

Hold the bottle over a bowl and fill it with water. Open each hole in turn. See how strongly or weakly the water comes out. Close each hole after you try it, and fill the bottle again. Lastly, open all the holes at once. Let the bottle drain empty. Watch the three jets of water and compare their strength. See if they change as the bottle empties.

❤ Record

Describe the experiment, and draw it. Think why the water came out when you opened the holes. Why did the jets have different strengths? Did the amount of water above each hole matter? Would the jets be stronger if the bottle were taller?

Picture 2 As water becomes deeper, pressure increases

Water pressure

Water is heavy. It presses downwards heavily. In your bottle, the water pressed against its bottom and sides. It pressed most strongly at the bottom, because the water was deepest there. It pressed less strongly at the top because the water was shallow there. We say that there was strong pressure at the bottom and weak pressure at the top. When you opened the holes, this pressure pushed the water out. It pushed it weakly at the top and strongly at the bottom.

The deep sea

The deeper you go under water, the more water there is above you. It is all pressing down. It is making pressure. Deep down, water pressure is tremendous. The captain in the submarine could not go deeper. The water would pour in more and more strongly. If submarines go too deep, water pressure can crush them.

Exploring

Men explore the deep sea in steel balls which are very strong. Men have been down nearly 11 kilometres to the deepest part of the sea. This is the Challenger Deep in the Marianas Trench, in the Pacific Ocean.

❤ *To write*

1 Why does deep water have stronger pressure than shallow water?
2 Why is it dangerous for submarines to go too deep?
3 What shape is strongest to stand up to pressure?

Picture 3   This diagram shows why pressure increases

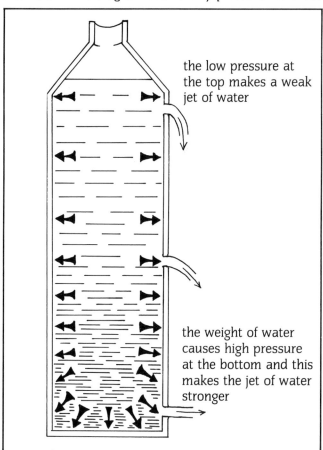

the low pressure at the top makes a weak jet of water

the weight of water causes high pressure at the bottom and this makes the jet of water stronger

Picture 4   This strongly-built submarine can go down to great depths

Unsafe to drink

Some water is not safe to drink. Why is this?

You will need water, six jars, sand, flour, salt, oil, sugar, sawdust, a stirrer, a saucer

♥ Experiment: To show how some things dissolve

Half-fill six jars with water. Take some things to mix with the water. Here are some ideas: sand, flour, salt, sugar, sawdust and oil. You can try other things as well. Choose one of them and tip a little into one of the jars of water. Stir it well and watch closely. See if it stays as it is, or disappears. If it disappears, it has dissolved in the water. Try all of your things, one by one. Use a different jar of water each time. See which things dissolve in water, and which do not.

♥ Record

Describe how you did the experiment. Make this chart to show your results:

What we put in the water	What happened
sand	
flour	
salt	
sugar	
sawdust	
oil	

Look at the jars with the things that did not dissolve. Could you get these things out of the water again if you poured the water through cloth? Try it. Look at the jars where things dissolved. Can you get these things back out of the water?

Picture 1 Set your experiment up like this

Pollution

We need to be careful about what we drink. For example, never drink straight from a river. Rivers can carry harmful things in them. When they do, we say that they are *polluted*. You can see some pollution. You can see bottles, cardboard, oil and junk. You can see clouds of mud being carried along in the water. Some pollution you cannot see. This is stuff that has dissolved in the water. Waste chemicals may pour into rivers. Fertilisers may be washed into streams. They dissolve in the water and cannot be seen. Some waste chemicals are poisonous.

Picture 2 Detergent foam is polluting this stream

Picture 3 This crop-spraying helicopter could pollute rivers accidentally

Distilled water

When water evaporates, it goes into the air as an invisible vapour. This vapour is made of pure water. Anything that is dissolved in the water gets left behind. If the water vapour is collected and turned back into water, this water is perfectly clean. We call it *distilled* water.

Evaporating water

You can prove that water vapour leaves dissolved things behind. Dissolve some salt in a saucer of water. Leave the saucer in sunlight for a few days. The water will evaporate, but the salt will be left behind in the saucer.

Picture 4 Why did the salt stay behind when the water disappeared?

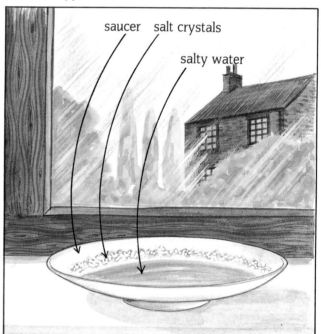

♥ To write

1 Why are some of our rivers 'dead' in the sense that no fish live in them?
2 Why is it a bad idea to drink from rivers?
3 Can polluted water be made clean again?

Changes in air and water

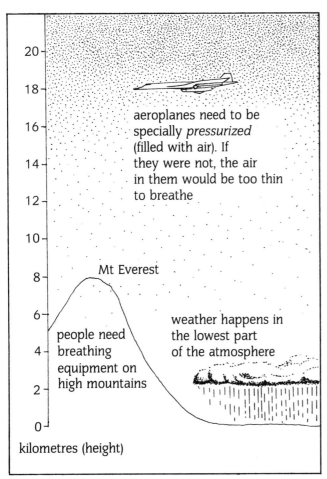

Picture 1   The Earth's atmosphere is made up of layers

The Earth is surrounded by a layer of air. This layer is many kilometres thick. It is called the *atmosphere*. The air is denser (thicker) near the ground. Its pressure is greatest at ground level. Higher up, the air gets thinner until it shades away into space. There is no air at all in space.

Burning

Things cannot burn very high up, because the air is so thin. There is not enough oxygen to feed the fire. Sometimes, aeroplanes can put out fires by climbing into thinner air.

Cold air

Air high up is colder than air near the ground. This is because air gets most of its warmth from the ground and the sea. Sunlight does not warm the air much as it shines through it. Many high mountains are covered in snow. People who climb mountains need warm clothes, even in summer.

Picture 2   Skiers have to dress warmly even when the sun is shining

Clouds

The air contains water vapour. When this vapour gets cold, it makes clouds. Clouds usually form high up, because the air is colder there.

The colour of the sky

The sky looks darker the higher you go. When you leave the atmosphere altogether, the sky is black. The sky only looks blue from the ground. This is because the air affects the sunlight shining through it.

The sea

The bottom of the sea can be up to 11 kilometres deep. The sea floor is different from the surface. It is as dark as black ink, and very cold. Sunlight does not reach it to lighten it or warm it. Most of the life in the sea lives near the light and warmth of the surface. But some plants and animals live even in the deep parts of the sea.

Diving deep

Deep down, the pressure of the water is tremendous. Human beings cannot dive deep without special help. The water would crush them. Men have made special submarines to hold out the pressure of deep water. They have used these to explore the bottom of the sea.

Picture 3   A diver exploring the sea's darkness

Sea water and rainwater

Sea water is salty water. Much sea water turns into water vapour. This happens at the surface of the sea. Later, the water vapour becomes cloud. In the end, it falls as rain. Yet rainwater has no salt in it. When sea water evaporates into the air, it leaves its salt behind. Evaporated water is pure water.

Picture 4   How sea water becomes rainwater and then returns to the sea

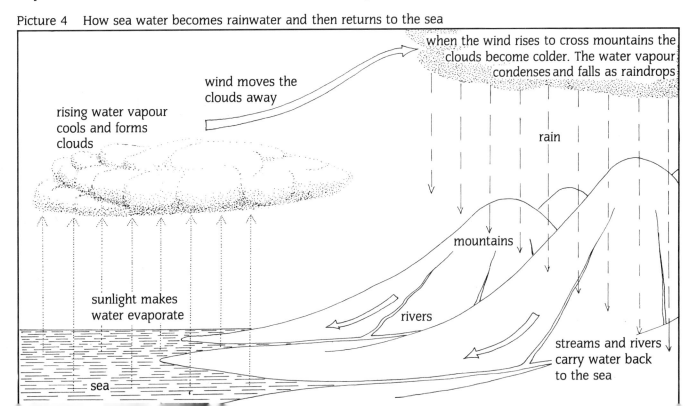

Part 3 Changes seen and unseen

Changing colours

Lots of quick changes puzzle our eyes. You can show this simply.

> You will need card, scissors, string, coloured felt pens, a pencil, plastic tubing, a magarine tub, masking tape

❤ Experiment: To make colours appear white

Cut a disc out of card, ten centimetres across. Mark the exact centre. Divide it up like Picture 1 shows. Colour each part the same as the picture. Use pieces of plastic tubing to fix the disc to the pencil. Make two holes in a margarine tub and slide the pencil through. Use more tubing to hold it in place. Fix a length of string to the pencil with a small piece of masking tape, and roll the pencil so that the string winds around it. Hold the tub firmly and pull the string. The disc will spin quickly. Watch the colours. They will look grey or white.

❤ Record

Explain what you have done. Describe the spinner disc and draw it. How would you describe the colour of the

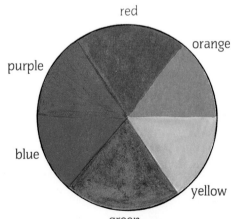

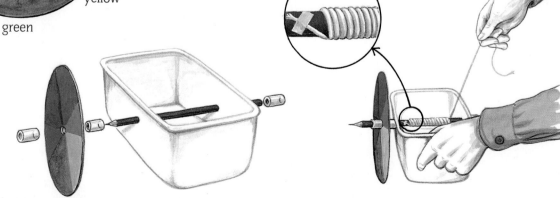

Picture 1   How to make colours look white

spinning disc? When the disc spun, did the sections really change colour? Why did they seem to change? Did it make a difference how fast the disc spun?

The spectrum

There is a name for the group of colours you used. It is called a *spectrum*. Isaac Newton first made a spectrum in 1666. He used a specially shaped piece of glass called a *prism*. He shone a ray of ordinary sunlight through it (we call ordinary light 'white' light). The light changed direction and split into different colours. It made a spectrum of light. White light is made up of light of different colours.

Why the disc looked white

Just as white light can split into colours, colours can mix to make white. Your spectrum disc fooled your eyes into

Picture 2   White light split into colours

violet
indigo
blue
green
yellow
orange
red

Picture 3   The colours of the spectrum

thinking that they saw white light. As the disc spun, your eyes saw red, violet, blue and so on quickly, one after another. Your eyes saw the colours mixed together. They saw them as white. (Your disc may have looked grey. This is because it is hard to get the colours just right. It is hard to spin the disc fast enough.)

❤ To write

1   When would you expect to see a natural spectrum?
2   How can you get coloured light from white light?
3   Your eyes can be tricked. True or false?

Picture 4   A rainbow is a natural spectrum

Changing what we see

Do our eyes always tell us the truth? Or do they sometimes 'change' what we see?

A simple illusion

Look steadily at Picture 1. What does it show? Keep on looking. After a while it will seem to change into something else.

The picture will not really have changed. The change you see is an *illusion*. An illusion is something that seems to be real but is not.

Picture 1

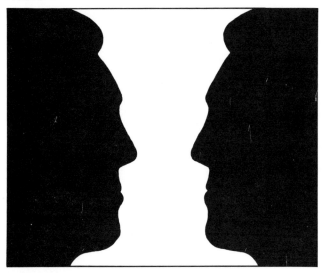

You will need paper, scissors, a stapler, a black felt pen

♥ *Experiment: Make an illusion*

Staple twenty squares of paper into a little book. On the first page, draw a circle in the bottom left hand corner. Draw another circle on the next page, slightly nearer the top right hand corner. Draw a circle on every page, and draw it nearer the top corner each time. The last page should have the circle in the top right hand corner. Then flick the pages. The circle will seem to move. You can try making more interesting illusions. Try drawing people and making them move.

♥ *Record*

Describe what happened when you looked at the picture. Say how you made a circle seem to move. Did the picture really change? Did the circle really move? Can

Picture 2 How cartoons are made

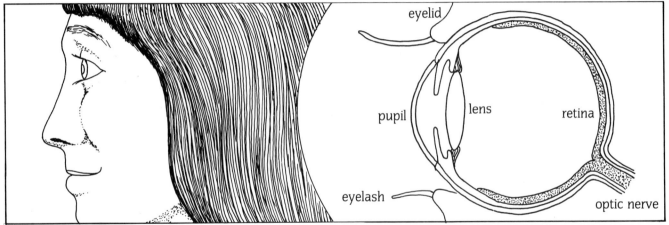

Picture 3   How the human eye works

our eyes sometimes 'see' what is not really there? Have you discovered how cartoon animals seem to move?

The eye

An eye has a special part called the *retina*. Whenever light goes into an eye, it touches the retina. The retina tells the brain when light touches it. It sends messages along a *nerve*.

Sorting the messages

The brain sorts the messages from the eyes, and arranges them into a pattern. This pattern is the 'picture' that we see. The brain tells us that there is a desk, or a table, or food, or a book in front of us.

Confusing messages

Sometimes the messages from the eyes can puzzle the brain. When you flicked the book, the messages showed the book and your hand in the same place. But they kept showing the circle in a different place. Your brain made the best sense it could of this puzzle. It decided that the circle must be moving. So you 'saw' the circle as moving.

Cartoons

Cartoons are made of thousands of pictures. Each picture is slightly different from the one before. They are shown one after another, quickly. This makes the pictures seem to move.

 To write

1   Where does light enter the eyes?
2   Can you say something about the speed at which a message is carried along a nerve?
3   How do cartoon films work?

Picture 4   Our eyes let us see the world around us

There or not there?

Picture 1   Make camouflaged 'animals' like these ones

Animals try to confuse the eyes of their enemies, and their prey. How do they do this?

You will need card, scissors, glue, black paper and coloured paper

❤ Experiment: Spot the target

Cut four large squares of white card. Leave one blank. Make patterns on the others by glueing on paper. The picture shows you what patterns to make. Then cut four smaller 'animals'. Make one to match each square.

Stand the squares at one end of the room. Stand someone as a hunter at the other end. He must cover his eyes. Arrange some 'animals' in front of squares that do not match them. The hunter must uncover his eyes for a moment. Then he must say how many 'animals' he saw. Then arrange the 'animals' differently, and let the hunter look again. Sometimes, put all of them out. Sometimes, put only one out. Sometimes, put them where they match; sometimes not.

❤ Record

Keep a record like this.

What was really there	What the hunter saw
2 animals matching the squares	one animal
3 animals not matching the squares	two animals
1 animal matching the square	
4 animals not matching the squares	

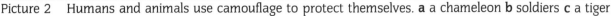

Picture 2   Humans and animals use camouflage to protect themselves. **a** a chameleon **b** soldiers **c** a tiger

Describe the game. Say if you managed to fool the hunter about the numbers of animals. What is the best way to fool him? Are some patterns better at hiding things than others?

Camouflage

Over millions of years, animals have changed. They have *evolved*. Most kinds of animal have changed the way they look. They have changed to match their surroundings as much as possible. When something matches its surroundings it is *camouflaged*. Your experiment showed that when something matches its surroundings it is harder to spot. The pictures show you some animals with their camouflage.

Changing camouflage

Some animals have gone a stage further. They have different camouflages for different times of the year. Some grow a special white coat for winter. This helps them to hide in the snow. Some fishes and lizards can do better still. Look at Picture 2. It shows a kind of lizard called a chameleon. It can change colour to help it hide in different surroundings.

Human and camouflage

People have copied the idea of camouflage from animals. Look at the way that soldiers use camouflage.

Eyes and camouflage

Camouflage works because animals' brains like to see patterns. When things are camouflaged, they make one pattern with their surroundings. The hunter's brain does not notice them. This is one reason why hunting animals need a good sense of smell. Their eyes can be too easily fooled.

❤ *To write*

1   It may be dangerous to us if we are not easy to see. Give an example.
2   Give an example of camouflage by shape.
3   How do hunting animals help their eyes?

A hidden change

Some changes are invisible. Electricity gives a good example.

Picture 1 Electricity can be dangerous

You will need a compass, a length of insulated copper wire, a jar, water, salt, blotting paper, pieces of zinc and copper

♥ *Experiment: To make electricity*

Wind some wire round a compass, in line with the needle. Leave the ends free and bare. Lay the compass flat on the desk. Soak some squares of blotting paper in a mixture of salt and water. Get ready six pieces of zinc and six pieces of copper. Arrange the copper, zinc and paper into a pile. The order should be: copper, paper, zinc, copper, paper, zinc, copper, paper and so on. Hold the pile firmly. Put one of the compass wires on one end of the pile. Put the other wire on the other end. Watch the compass needle. It will jerk. The wire around it has changed invisibly. This change affects the compass. The wire has got electricity in it.

Picture 2 How to make electricity

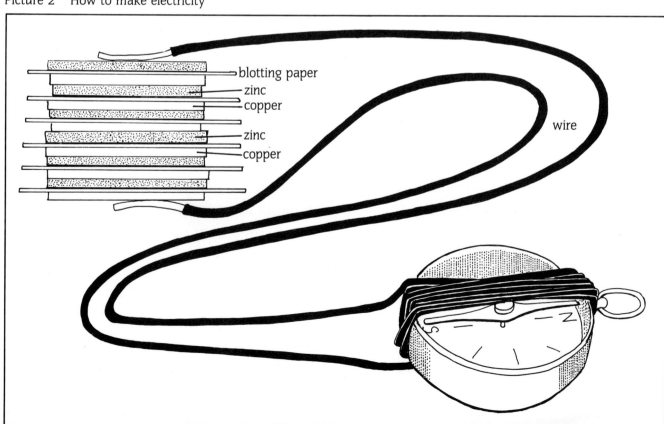

♥ Record

Describe the experiment. Draw it. Say how you know that a change happened in the wire. What caused the change? Could the change happen without the pile of metals? Must the metals be different from each other? Does it matter what metals you use?

A voltaic pile

You have made a copy of the first battery ever made. Alessandro Volta made a pile like yours about two hundred years ago. It is called a 'voltaic pile'. Volta realised that we can make electricity by putting certain metals together in certain ways.

A voltaic cell

An improvement on the voltaic pile was the voltaic cell. In a voltaic cell, plates of copper and zinc are put into acid. The copper and zinc plates have a special name. They are called *electrodes*. The acid is called the *electrolyte*.

A dry cell

Modern batteries are made of dry cells. The electrodes in the dry cell are made of zinc and carbon. There is a special paste between the electrodes, which is the electrolyte. The paste is much safer than acid. In a cell, the electrodes and electrolyte gradually change. This change makes electricity. Batteries do not make electricity all the time. They only make electricity when they are joined to things that can use it.

♥ To write

1 What electrical measurement is named after Alessandro Volta?
2 Why is electrolyte paste safer than acid?
3 How can you make batteries start to produce electricity?

Picture 3   Modern batteries come in lots of shapes and sizes, and can supply different amounts of power

Picture 4   This milk-float runs off batteries

Let there be light

Why does a bulb in a torch light up? If you touch a bulb to a battery, nothing happens.

 You will need a battery, a bulb, two lengths of wire

 Experiment To show how electricity works

Touch a bulb to a battery. Nothing will happen. Now take a wire with bare ends. Use the wire to join the metal part of the bulb to the bottom of the battery. The bulb will light. Take a second wire. Arrange the wires, bulb and battery as Picture 1 shows. One wire must join the top of the battery to the end of the bulb. The other wire must join the bottom of the battery to the metal part of the bulb. The bulb will light.

See if any other arrangement will make it light. For example, try it with both wires touching the same end of the battery.

 Record

Describe how you joined the bulb and the battery to make the bulb light. Draw how you did it. Say what other ways you tried, and what happened. Will a battery light a bulb only if they are joined in a special way? Will a battery make electricity only if things are joined like this?

Picture 1   Arrange the batteries and bulb like this

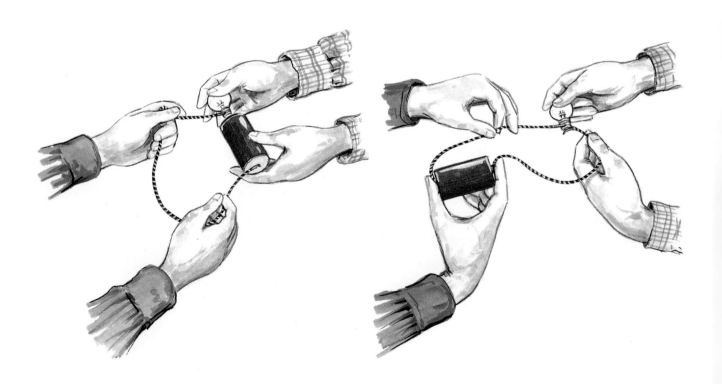

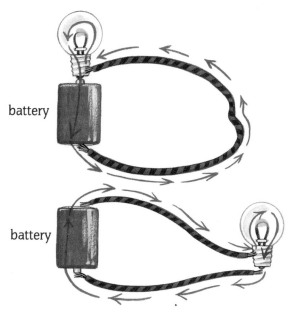

Picture 2 This is why the bulb lights up

Circuits

Electricity lights a bulb by going through it. It does not go into the bulb and stay there to be used. It needs to go through the bulb and back to the battery. If it cannot do this, it will not light the bulb.

An electrical circuit

Electricity needs to travel in a kind of circle, or *circuit*. We usually make circuits out of wire. Electricity moves easily through wire. Your bulb would light only when you put a wire between the bulb and the battery. This wire lets the electricity move through the bulb and back to the battery. It made a circuit. You made another circuit using two wires.

How a torch works

In a torch, one end of a battery touches the bulb. Metal goes from the other end of the battery to near the bulb. When you turn on the torch, the metal is made to touch the bulb. This makes a circuit. Electricity travels round it and lights the bulb.

Picture 3 Torches can be very useful!

❤ To write

1 Why does electricity need a circuit?
2 Why do we usually use wire in circuits?
3 When you turn on a torch, a circuit is completed and the bulb lights up. How does the switch complete the circuit? (Unscrew the torch to find out.)

Picture 4 How a torch works

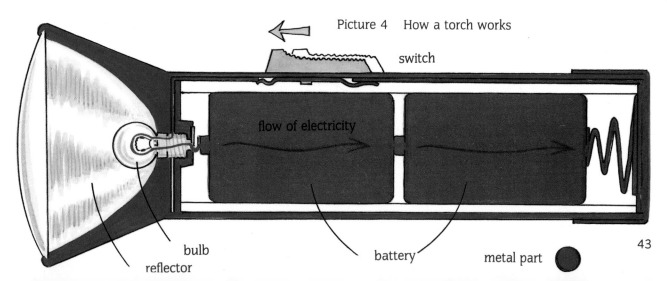

Invisible change

Collect some different things together including some made of metal. Hold a magnet by each one in turn. You can see that some things are affected by the magnet. Some are not. You can show this in an even clearer way.

 You will need a large iron nail, a piece of wood, a piece of plastic, a piece of card, iron filings, a bar magnet

♥ Experiment: To magnetise a piece of iron

Pour some iron filings on a sheet of paper. Take an iron nail and your magnet. Use the magnet to stroke the nail twenty or thirty times. You must stroke one way only, with the same end of the magnet. Lift the magnet well away from the nail after each stroke. Then put the nail by the filings. See if it pulls them. If it does, you have made a magnet. Try stroking wood, card and plastic like this. See if they become magnets. Try other things.

Picture 1   How to turn a nail into a magnet

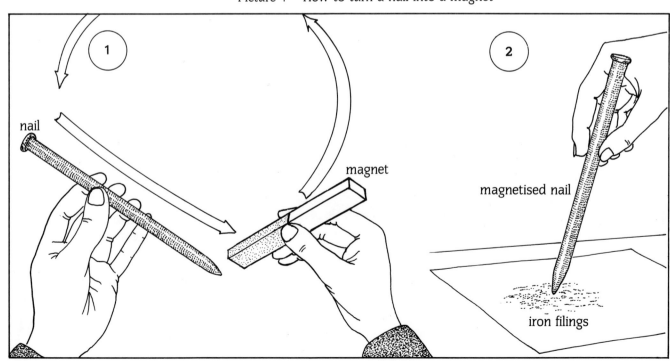

 Record

Describe how you changed the nail into a magnet. Say how you know it was a magnet. Draw a picture of the experiment. Say how you tried to *magnetise* wood, card and plastic. Can you make a magnet out of anything? What kinds of things are affected by magnetism? Does something look different when it becomes a magnet? How does it change?

What magnetism is

Magnetism is an invisible force which pulls at some things. We cannot see or feel magnetism, but we can see what it does. Magnetism is not the only invisible force. Electricity is invisible. So is gravity, an invisible force holding us to the ground. Magnets do not look special from the outside. If you cut a magnet in half, the inside looks ordinary.

The magnetic field

Your iron filings may have jumped to your nail before it touched them. If you put the magnet near them, you will see them move. This shows that magnetism reaches outside the metal of the magnet. It makes an area of magnetism round the magnet. We call this the *magnetic field*.

Picture 2   A diagram of a magnetic field

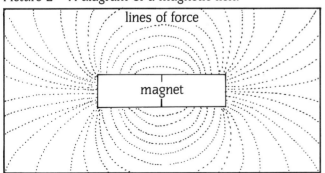

Magnetic and non-magnetic

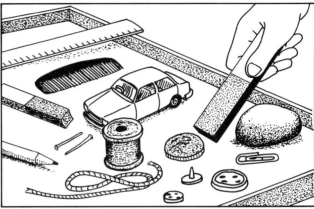

Picture 3   Test your magnet on different things

Not all things are affected by magnetism. You found this with wood, card and plastic. These things are not pulled by magnets. They cannot be made into magnets. They are non-magnetic. Only a few metals are magnetic. The best-known are iron and steel.

Picture 4   This huge magnet is used to lift very heavy objects

To write

1   How wide is the magnetic field of your magnetised nail? Can you make it larger?
2   Did you find any non-metal that could be magnetised?
3   How is magnetism like electricity and gravity?

Sensing change

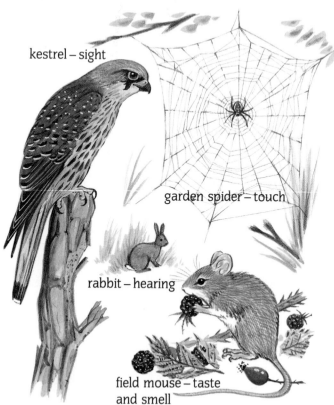

Picture 1 Animals' senses tell them what is happening

Our senses tell us what the world around us is like. They warn us when something changes. For example, they warn us when it gets hot or cold. They warn us when things suddenly move.

Where are our senses?

Our bodies have special parts called *receptors*, which can tell when changes happen. They are sensitive to change. Some are sensitive to one kind of change. Some are sensitive to other kinds. Our skin has parts that are sensitive to touch. Other parts are sensitive to heat and cold. Our eyes have parts that are sensitive to light.

Picture 2 Nerves carry messages very quickly from one part of the body to another

People are part of the animal kingdom. Like all animals, we need to know what is happening around us. Our bodies have senses that do this.

Our senses

Our senses are things which send messages to our brains. We have five kinds of sense. The sense of touch tells when something touches us, or we touch something. The sense of sight tells us what things are further away. The sense of taste tells us what is in our mouths. The sense of smell tells us what is around us that we cannot see. The sense of hearing tells us what is going on that we may not be able to touch, see, smell or taste.

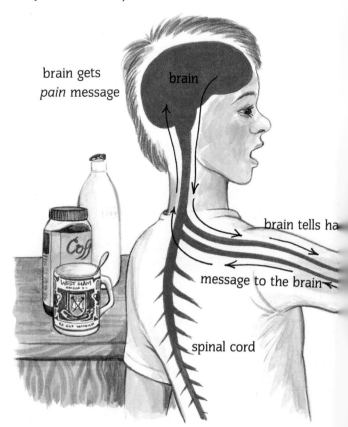

Nerves

All these parts are at the ends of nerves. Nerves connect the receptors to the brain. If a hand touches something hot, for example, it sends a tiny amount of electricity along a nerve. The electricity travels to the brain. It is a message which tells the brain about the change of heat. The brain decides what to do. It might decide to move the hand. The brain sends messages back, telling the arm muscles to move the hand. All this can happen at great speed.

The nervous system

A body has thousands of nerves in it, all connected together in one system. This is the *nervous system*. Picture 3 shows how nerves connect every part of the body with the brain.

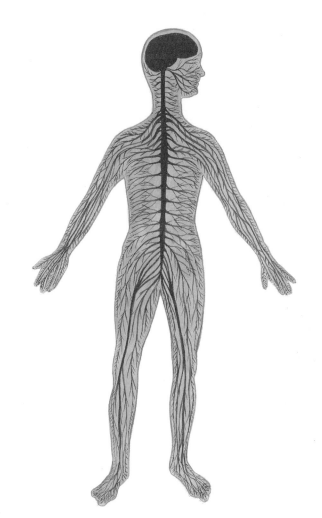

Picture 3 A simple drawing of the human nervous system

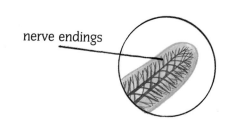

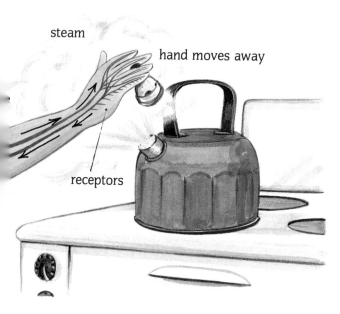

Picture 4 What do this boy's sense tell him?

New words

atmosphere   the Earth's covering of air

bacteria   tiny living things that help to decompose dead things

camouflage   an arrangement of colour or shape meant to fool the eyes

chlorophyll   the green stuff in plants that lets them make starch from sunlight, air and water

circuit   a pathway for electricity

climate   the weather over a long period

condense   to turn from vapour into a liquid

decompose   to rot

decomposer   an animal or plant which helps things rot

dissolve   to mix with a liquid

distilled water   pure water made from water vapour

electrode   the solid parts inside a battery which make electricity

electrolyte   the part of a battery which surrounds the electrodes

equator   an imaginary line around the world exactly half way between the north and south poles

evaporate   to turn from a liquid into a vapour

evolve   to change over millions of years

fungus   a plant that does not have chlorophyll

germinate   to begin to grow from a seed

hibernate   to sleep through winter

humus   the stuff made when plant and animal remains rot

illusion   something which seems real, but is not

loam   soil that is good for growing plants

magnetic field   the invisible area of magnetism around a magnet

magnetise   to turn into a magnet

migrate   to go to a different place for a season

nerve   a thing like a thread which carries messages to and from the brain

nervous system   the whole collection of nerves in the body, including the brain

oxygen   part of the air that we breathe

polluted   spoiled or made unhealthy

pollution   anything that spoils something or makes it unhealthy

pressure   a steady push, usually made by the weight of something pressing downwards

prism   a specially shaped piece of glass or plastic

receptor   a part of the body that senses change

retina   part of the eye that is sensitive to light

senses   the things our bodies use to tell our brains about the world

water vapour   water which is invisible in air